CRACKING THE PATENT CODE

Create A Competitive Edge and Protect
Your Inventions From Being Stolen

STEVEN SPONSELLER
INNOVATION STRATEGIES, INC.

Copyright © 2020
Steven Sponseller
All rights reserved.

No part of this publication may be reproduced, distributed, or transmitted in any form or by any means, including photocopying, recording, or other electronic or mechanical methods, without the prior written permission of the publisher, except in the case of brief quotations embodied in critical reviews and certain other non-commercial uses permitted by copyright law.

Innovation Strategies, Inc.
624 W. Hastings Road, Suite 3
Spokane, WA 99218
www.SteveSponseller.com

First Edition 2020

ISBN: 979-8682228263

This book is dedicated to the most important people in my life. My wife, Beth, and my daughters, Mia and Zoe.

CLIENT TESTIMONIALS

"After working with numerous attorneys, I was fortunate to be introduced to Steve Sponseller. What a difference. Steve became an integral team member of our technical staff. He does his homework and possesses the ability to quickly interpret a specific technology or invention in terms of intellectual property value."

Marketta Silvera

Silicon Valley CEO/Chairman

"Steve is an excellent intellectual property lawyer. We've worked with Steve over the past 20 years and continue to work with him. He is a quick study. In the course of just one meeting, he understands the core concepts of what we are trying to protect. Steve is easy and fun to work with!"

Dilip Venkatachari

Board Director, SageX Solutions

"Steve's training helped me understand what I needed to do to protect the inventions in my business. Steve's expertise allows him to present ideas in a very easy-to-understand way that anyone can execute. Based on the training, I am confident that the intellectual property in my business is now protected."

Ken Lovett

CEO, Neuroprima Inc.

TABLE OF CONTENTS

INTRODUCTION 1
CHAPTER 1 7
   Elevate the Importance of Patents In Your Company
CHAPTER 2 17
   Create a Steady Flow of New Inventions
CHAPTER 3 25
   Be Proactive and Find Your Gold Nuggets
CHAPTER 4 35
   Evaluate Inventions Using Criteria Specific to Your Company
CHAPTER 5 45
   Protect the Right Inventions
CHAPTER 6 55
   Build a Patent Portfolio Filled with High-Value Patents
CHAPTER 7 63
   Patent Boosters
CHAPTER 8 71
   Critical Events that Require Patent Action
CHAPTER 9 85
   Patent Pitfalls to Avoid
CONCLUSION 93
ABOUT THE AUTHOR 95
   Who I Am
NEXT STEPS 97
RESOURCES 99

INTRODUCTION

"The day before something is a breakthrough, it's a crazy idea." Peter Diamandis

I've gone from building toy spaceships as a kid to working with cutting-edge inventions that have a global impact. I have a passion for learning about the latest innovations and love working with technology companies to discover and protect the latest inventions.

Instead of toy spaceships, today I enjoy working with the latest computer systems, autonomous vehicles, data management systems, and artificial intelligence algorithms. I feel lucky because I get to see the latest inventions before the rest of the world!

This book is the result of my 20 years of experience working with hundreds of technology companies and more than 1000 inventors. I am excited to share this information with you.

Patents Are Critical for Every Technology Company

You invest significant time, money, and effort developing new products and services. That means lots of late nights and missed family activities to perfect these inventions. You and your team are doing important work. Your inventions can make vehicles safer, get vaccines to patients faster, and provide improved teaching systems for kids.

You want to protect those inventions just like you would any other valuable business asset. You may not be able to put inventions under lock and key, but you can file a patent application. Patents provide a way to protect the inventions in your new products and services.

Patents are vital to your company because they provide a competitive advantage and protect your inventions from being stolen. For example, your inventions can disrupt your industry and position the company as a market leader. Patents covering critical inventions in your industry provide a competitive edge during the life of the patent.

New competitors can pop up overnight. Without strong patent protection for your core technology and must-have product features, you may not be able to stop competitors from stealing your market share with copycat products. A well-designed portfolio of patents can prevent theft of your sales by these competitors.

A Three-Step System to Identify and Protect Your Most Valuable Inventions

In my experience, many companies do not understand the patent process and fail to properly protect inventions that are critical to the company's success. That's why I created this system – and it's a proven system.

The three-step system I describe has been tested with hundreds of companies of all sizes. It has worked with many technologies, including computer systems, autonomous vehicles, data processing systems, artificial intelligence, and robotics, just to name a few.

My system has three important steps that produce a valuable portfolio of patents. It's important that you don't skip any of them because all three steps work together to identify and protect only the most valuable inventions. You don't want a collection of mediocre patents. If your patents fail to cover important inventions, they are of little value to your organization. Weak patents don't provide any real protection.

The three-step system I describe also helps you discover a gold mine of inventions hidden throughout your team. You have a team of smart people with fantastic ideas. There are ways to harness the collective innovation of your team to create valuable inventions. I will show you how.

I am sharing my system with you to clarify the invention process and my system for protecting your inventions using carefully selected patent applications. This is the same system I use every day with my own clients. Each time I use this system with a new client, they are amazed at its simplicity and effectiveness.

When you implement this system, you will create a strong and secure portfolio of patents that positions your company as an industry leader. As your patent portfolio grows, you will enjoy increased protection from competitors and a higher company valuation. Remember, an expanding patent portfolio affords peace of mind knowing that your core technology and must-have features are protected from being ripped off by copycats. This system will change the way you think about your inventions.

No More Mediocre Patents

If you want to learn how to identify and protect the critical inventions developed in your business, you're in the right place. If you are confused about the patent application process and don't understand which inventions are worthy of patent protection, this book sheds light on those issues.

Don't worry, I won't overwhelm you with a complicated 47-step system.

Instead, this book describes three critical steps to identify and secure your most valuable inventions. You can start working on these steps today.

If you work with a technology company of any size, these steps will help you and your team create a strong patent portfolio that provides a competitive advantage and protects your critical assets from being stolen.

Are you ready to start building (or expanding) your patent portfolio?

Keep reading to get started.

STEVEN SPONSELLER

CHAPTER 1

ELEVATE THE IMPORTANCE OF PATENTS IN YOUR COMPANY

A system that helps you to rate
Your inventions, from good to great
To show what's essential
To your company's potential
And stay in a competitive state
  – Cliff Feightner

I started working with Stark Enterprises when they had one issued patent and no pending patent applications. To say that Stark was exposed is an understatement. Although Stark had over 600 employees and over $200 million in sales, their patent portfolio was microscopic. The company was "naked" with respect to their patent protection.

Note: I changed the company name to protect the actual parties discussed in this case study.

Stark was a fast-growing company in the data communication industry. It had developed new systems and algorithms to communicate data to distributed locations in a highly efficient manner. Stark's systems and algorithms were much faster than its competitors and it was taking market share from others in its industry.

But Stark was at risk of losing its competitive advantage because it had not adequately protected its inventions, such as the unique systems and algorithms that improved data communication efficiency. By failing to protect those inventions, any competitor could freely copy those unprotected inventions and use the unique systems and algorithms in their own products.

And that's exactly what happened.

One day I received a frantic phone call from Katherine, the CEO of Stark Enterprises. She was distressed because a competitor, Leftfield Corporation, had just launched a new product that used systems and algorithms very similar to those created by Stark. Leftfield's new product took away Stark's competitive advantage.

Although I was working with Stark to build its patent portfolio, and there were many pending patent applications, the company had no issued patents that could exclude Leftfield from using competing systems or algorithms. Stark's only issued patent was focused on an invention that had no value to the company and did not cover the competitors. So, Stark had no way to stop Leftfield's new product.

As a result, the company's growth slowed because Leftfield's new product cut into Stark's sales.

This was a wake-up call to Stark's management team.

Fortunately, we had already started to build a patent portfolio that would help the company avoid this situation in the future.

The company set a goal to build a strong patent portfolio, and I worked with the Stark team, using the same system described in this book. We didn't focus on mediocre inventions that offered no real protection. Instead, we implemented systems to create industry-leading inventions and build a patent portfolio filled with high-value patents.

The company is now in a much better position to defend against competitors who attempt to copy their proprietary technology. Stark's portfolio now has numerous issued patents and many pending patent applications covering a wide variety of inventions. The patent portfolio continues to grow as the company develops new inventions that are stimulated by the growing innovation culture throughout the organization.

Katherine and the executive team now have peace of mind knowing that their systems are automatically creating new inventions from team members throughout the organization. Stark is stronger because the company has a growing patent portfolio that protects the company today and is expanding to provide even more market-leading assets in the future.

Just like Stark Enterprises, your company makes substantial investments of time and money to develop new products and services. The inventions in those products and services are critical to the growth of your organization. So, you want to protect those inventions as you would any other business investment or business asset.

Patents provide this protection.

Unfortunately, many technology companies take a haphazard approach to protecting their inventions. They have no system to identify inventions, prioritize those inventions, and build a portfolio of patents that provides protection and significant value to the company. Too often, I've seen companies randomly file patent applications without performing any type of evaluation of the invention being protected. This haphazard approach doesn't work and is often a waste of time and money. For example, spending $20,000 or more for an issued patent that covers an obsolete invention is a bad idea.

Why Patents are Critical to Technology Companies

A well-defined and carefully crafted patent portfolio provides multiple benefits for your technology company. Your patent portfolio is a collection of enforceable patents that protect high-value inventions.

1. *Competitive advantage.* A patent portfolio can protect your company's most vital assets – inventions that distinguish your organization in the marketplace. Technology companies are often founded on technological inventions that improve upon existing products. This improved technology and future inventions can give your company a competitive edge if it's protected from being copied. Obtaining patent protection on these critical inventions provides a barrier to entry for competitors. If your company has patents protecting must-have features in your industry, those patents provide a strong competitive advantage.

2. *Stop copycats.* If your company's inventions become popular in your industry, competitors will try to copy those inventions. Strong patents can stop those competitors from copying your inventions and may provide an additional revenue stream if you license certain patents to other companies.

3. *Attract investments (and get favorable terms).* Savvy investors want to see that your company is taking steps to protect its intellectual property assets. Investors frequently ask what has been done to protect proprietary technologies. A portfolio of strategic patents (and patent applications) provides proof that your organization has protected its important business assets. A portfolio of patents that provides broad protection can improve your company's valuation, thereby improving your position when negotiating investment terms. Your patent portfolio can give you an edge over other companies that have weak or nonexistent patent protection and are less attractive to investors.

4. *Discourage aggressive competitors.* As your company grows and enjoys an increased market share, larger competitors may take notice and want to reclaim their lost sales. To retaliate, these competitors may initiate a patent infringement suit against your company. Without a valuable patent portfolio, you may be targeted due to the lack of patents to fight back against the competitor. However, a strong patent portfolio may discourage the competitor from initiating legal action. And, if your company is sued for patent infringement, a strong patent portfolio puts your company in a position to negotiate a better settlement and reduce disruption to business operations.

You Want a System to Identify and Protect the Most Valuable Inventions

I've seen dozens of tech companies take a sloppy approach to identifying and protecting inventions. These companies obtained a few patents, but the patents failed to adequately protect the company's most valuable inventions.

For example, I've reviewed patents for companies that protected a few random inventions. But those inventions were not the critical core features that distinguished the company's product in the marketplace. Instead, the patents covered very specific implementation details that had become outdated. As a result, those patents could not stop competitors from using the company's important features. Essentially, the company had left the door wide open for any number of competitors to freely copy the company's most critical inventions.

Now let's talk about how you can identify and protect your most valuable inventions with my three-step system. I developed this system based on more than 20 years of experience helping technology companies identify and protect their critical inventions. This is the exact system I use every day with my own clients.

This system has three important steps that produce a valuable portfolio of patents. I have discovered that too many companies skip the first two steps and end up protecting mediocre inventions that provide little value to the company. If you ignore the first two steps of the system, your company will likely waste money on low-value patents and not obtain the full protection it deserves. These mediocre patents often provide a false sense of security because they don't protect your inventions from being copied.

Why three steps? It's simple to implement and focuses on the key activities. Besides, if I created a 27-step system, I would have scared you away from this book.

I enjoy taking complex systems and distilling them down to the most important parts. That's what I've done here. I have stripped away the complexity and confusion of identifying and protecting inventions, leaving a streamlined yet powerful system.

Here's a peek at the system.

Three-Step System

1. *Identify your inventions.* Identifying all of your organization's inventions is critical to building a strong patent portfolio. You can't take steps to evaluate and protect your inventions until you identify them in the first place. Chapters 2 and 3 discuss techniques for teaching your team members how to identify inventions, implement invention tracking

systems, and proactively identify your most valuable inventions. These chapters teach you how to create a steady flow of new inventions and make sure nothing slips through the cracks.

2. *Evaluate all inventions.* If you really want a strong patent portfolio, it's important to properly evaluate all of your identified inventions. Too many companies ignore this important step and end up preparing patent applications that do not provide maximum value. The strength and quality of your patent portfolio depends on a thorough evaluation. Chapter 4 describes how to create a diverse invention review committee, define appropriate evaluation criteria that support your company's goals, and determine the value of potential patent protection for specific inventions.

3. *Protect the right inventions.* To truly protect your company, you must build a patent portfolio with the most valuable inventions. Protecting the wrong inventions is a waste of time and money. Patents covering the wrong inventions are dangerous because they give a false sense of security. For example, a patent portfolio with many patents may seem valuable, but a portfolio of weak patents is not likely to provide the protection and value you believe it contains.

Chapter 5 discusses the key categories of inventions that must be protected and various techniques for protecting those critical inventions.

Later chapters will discuss each of these three steps in more detail and show you how to implement each step in your own business.

As you can see, the last step in the system is protecting the right inventions with a patent application. If you don't complete the first two steps (identifying and evaluating inventions), you're probably missing the most valuable inventions and, therefore, the most valuable patents.

The rest of this book will show you how to implement the three-step system and create a valuable patent portfolio that protects your most valuable assets. In the next chapter, we'll look at how to create a pipeline of new inventions that captures all of your team's creative ideas.

CHAPTER 2
CREATE A STEADY FLOW OF NEW INVENTIONS

"You can't use up creativity. The more you use, the more you have." Maya Angelou

In the previous chapter I discussed the importance of patents to technology companies and outlined my three-step system to identify, evaluate, and protect your most valuable inventions. Now, let's dive into the identify step, and I'll show you how to create a steady stream of inventions.

Many technology companies are founded on a core invention that significantly improves upon existing products or services. However, it's easy for these companies to become one-hit-wonders if they fail to continue innovating to improve their initial invention. Many tech companies fail because they stop the innovation process that created the business success in the first place.

Avoid the one-hit-wonder trap by taking steps to encourage new ideas and build a steady flow of new inventions. This chapter discusses the basic foundation for identifying and tracking inventions. The next chapter will discuss multiple activities to boost your team's creative thinking and identify more inventions, using contests, brainstorming sessions, and invention mining.

Let's start setting the foundation for identifying and tracking all of your inventions. After all, you can't take steps to protect your inventions until they are identified. It's vital.

Do this, and you'll thank me.

Teach Team Members How to Identify Inventions

Most business leaders and technical team members never learn about the invention process because it's not taught in college or high school. When I was an electrical engineer for a technology startup, I didn't understand how to identify inventions – and neither did my peers nor the company CEO. There was one guy named Don, who worked in the optics lab. It was Don's job to do the inventing. Leaving out the rest of the team was a big mistake!

Don't overlook the importance of this chapter and the need to teach your team how to discover inventions.

Building a strong patent portfolio requires regularly monitoring activities to be certain your organization is identifying and protecting the most valuable inventions. An invention is defined by the Cambridge Dictionary as "something that has never been made before, or the process of creating something that has never been made before."

In some cases, an invention is a revolutionary new product that disrupts an industry. But most inventions are incremental improvements of an existing system or process. For example, an invention may improve the speed of a microprocessor, using a new technique for communicating information between cores in the microprocessor. The improved microprocessor speed is valuable because the microprocessor can perform operations faster than the previous version. The new communication technique is an important invention worthy of patent protection since it provides a competitive advantage (faster microprocessor speed).

Don't assume that your team members understand how to identify inventions in their work. Instead, take the time to explain how to identify inventions and provide examples of inventions created within your organization.

In my experience, many developers and engineers overestimate the threshold for a patentable invention. The US Patent and Trademark Office has issued more than 10 million patents. The vast majority of these patents cover small improvements. Thus, the threshold for obtaining a patent is much lower than what most people believe. Even though the patents cover small improvements, those patents can be extremely valuable if they provide a barrier to entry for competitors. Show your team examples of patented inventions in your industry that represent small improvements yet provide significant value to the patent owner. These examples will help your team members identify more inventions and avoid disregarding valuable incremental improvements.

Identify Inventions as Part of the Product Development Process

One way to automate the process of identifying inventions is to incorporate invention discussions into the product development process. When developing a new product, the development team often creates inventions. For all new product projects, add at least one activity that requires team members to identify inventions.

It's best to identify these inventions when the project is still fresh in the minds of the team members. Some of my clients add an "invention identification" activity to all project development checklists. The project is not finished until that item on the checklist is complete.

When identifying inventions in a project, have the team members think about the biggest obstacles they had to overcome to complete the project. Often, the solutions to these obstacles are inventions.

For example, assume that the project is a new control system for a robot. In this case, the team faced the obstacle of obtaining sensor data related to objects located near the robot. The team considered several existing (i.e., off-the-shelf) sensors for use with the new control system. However, the existing sensors did not generate enough data to implement the advanced features of the control system. So, the development team designed a new sensor that generated data of a higher resolution that allowed the control system to operate more effectively.

Some team members might be thinking, "So what? It's just another sensor." But that sensor is likely an invention. Since there were no existing sensors that could generate the desired level of detail, the team had to invent a new sensor. This type of invention is important to consider for patent protection because it fills a gap in the marketplace. If the new robot control system needed this improved sensor to operate effectively, other products in the marketplace may need a similar sensor. It's important for the company to consider protecting their investment in this proprietary technology.

Keep the Invention Disclosure Process Simple

To identify the inventions developed by your team, make the process of disclosing those inventions simple. I've seen companies use an invention disclosure form with multiple pages and dozens of questions. I've also heard the development team groan about filling out the form. Many team members won't submit ideas because it's too time consuming. "I don't have 30 minutes to fill out the form for every idea" is a common complaint.

Instead of an overly-complex form, keep it simple. It's better to have a short invention disclosure form that actually gets completed than a complex form that is ignored. The goal is to get your team members to submit all of their inventions, so simple is better.

An invention disclosure form only needs a few components. There are other details needed if the company decides to pursue a patent for the invention, but those details can be collected at a later time.

Your invention disclosure form should include:

- invention title
- brief description of the invention (including the novel aspects of the invention)
- products/systems that use the invention

- names of inventors
- date the invention was created
- dates of any past public disclosure of the invention
- dates of any future public invention disclosure

Visit SteveSponseller.com/tools: to get an Invention Disclosure Form template you can use in your own company.

By keeping the invention disclosure form simple, more inventions will be disclosed so great ideas don't slip through the cracks.

Now that you have a simple invention disclosure form that's not a burden to complete, you want to track all received forms. This can be done with a database or any other tool that records and tracks the information on every form.

Your tracking system should also track the status of each invention disclosure form. This status information is important to keep track of how each invention is being handled. Each invention disclosure will be discussed by an Invention Evaluation Committee (as discussed in future chapters), which will decide how to proceed with the invention. The status of an invention may change over time. For example, your invention tracking system should track the following states:

- invention disclosure form received
- committee decision: do not pursue
- committee decision: prepare patent application
- committee decision: defer to future date
- committee action: request additional information from inventors
- committee action: request analysis by other team members
- committee action: schedule brainstorming session with team to discuss invention further

Depending on your own system for handling and reviewing invention disclosure forms, your invention tracking system may have additional or different states.

Now that you have a system in place to capture and track all inventions created by your team, the next chapter will show you how to proactively find inventions hidden throughout your organization. You won't just rely on team members to submit invention disclosure forms. You will take steps to stimulate more inventions and discover inventions that team members have not yet realized.

CHAPTER 3
BE PROACTIVE AND FIND YOUR GOLD NUGGETS

"The best way to predict the future is to invent it." Alan Kay

You may be sitting on a gold mine of inventions and not even realize it!

In the previous chapter, I discussed how to create a steady flow of new inventions and track those inventions. Now, it's time to leverage the knowledge and creativity of your team members to identify and create valuable inventions. In this chapter, I will show you several activities to help identify inventions hidden throughout your team. These inventions will jump-start the growth of your patent portfolio.

Get the Entire Team Involved

After working with hundreds of technology companies, I'm convinced inventions come from all areas of an organization. It's not just the engineers and the development team that create inventions. Every person in an organization has unique experiences and knowledge that provide a distinct perspective on the company's products or services.

People with different roles in the organization have different levels of interaction with customers and prospects. For example, people in the sales department regularly talk to customers one-on-one. These conversations can identify problems with the company products that are experienced by the customers. These problems can be reported to the development team, which can create solutions to the problems that may include inventions that can be patented. These are high-value inventions.

Additionally, sales team members may hear wish-list items from customers, such as features they wish the product had. These desired features can be created by the development team and added to the product to improve its appeal to current and prospective customers. If these new features include inventions, they may be protected to strengthen the company's portfolio of patents.

Here's one way it can work – and fast. Several years ago, I worked with a company that actively encouraged the sales team to get feedback from existing customers regarding ways to improve the company's products. This feedback was quickly evaluated and provided to the development team. The company leaders worked with the development team to revise the products to better meet the needs of the customers. This approach resulted in several new inventions that improved product sales and strengthened the company's competitive edge in their market. The customers were happy, and the company was more profitable.

Your customer service department can discover a variety of problems through customer service complaints and customer feedback. Problems reported by one customer are likely to be experienced by many other customers too. Solving these problems improves customer satisfaction and produces new inventions simultaneously. It's a win-win situation.

People in the marketing department typically monitor industry trends and customer buying habits. These trends allow the marketing team to identify new product opportunities and new features for existing products. By spotting these trends early, your organization can be first-to-market with a solution and may be able to patent that solution. For example, many companies are taking advantage of the growing use of artificial intelligence. These innovative companies are leveraging the increasing use of artificial intelligence algorithms to provide new and personalized features for their customers.

Tap into the collective creativity of your entire team by encouraging everyone in your organization to share innovative ideas with company leaders. And schedule events that encourage all team members to identify new ideas to improve existing products and create innovative ideas that can be developed into new products. I will discuss several of these events in Chapter 7 that can benefit your company by expanding your portfolio of valuable patents.

Invention Mining

How horrible would it be to invent something new and not get value from it?

One of the first activities I perform with new clients is called *invention mining*. This is a process of analyzing current products (and products approaching their release date) to be sure all critical inventions have been identified and protected. Typically, I meet with a client in person or via video conference and go through a list of questions to be sure they have identified all inventions. Using this process, companies often find gold nuggets of invention that were previously undiscovered.

The invention mining process includes meetings with the product development team to identify innovations in current products. It's important to evaluate the importance of these innovative systems and, if appropriate, protect these systems by filing a patent application. This evaluation is discussed in Chapter 4.

Invention mining is not limited to the product development team. Other departments in the organization can also help identify critical features of a product. For example, the sales team can identify particular product features that customers are most excited about and help the sales representatives sell the product. These exciting features are important to protect because a patent may allow you to exclude competitors from using these features. This is a valuable competitive advantage.

It's important to perform this invention mining as soon as possible because there are deadlines for filing patent applications, as prescribed by the US Patent and Trademark Office. Generally, if a product that contains an invention is sold (or offered for sale) the organization has 12 months from the date of first sale (or first offer for sale) to file a patent application covering the invention. If an application is not filed in the United States within 12 months of the first sale, you may lose out on protecting your invention.

Also, if an invention has already been disclosed in public, it is important to file a patent application as quickly as possible to avoid having another person file a patent application before your company does. The United States patent system is a "first inventor to file system." The "first inventor to file" system means that if two inventors develop the same invention, the first inventor to file their patent application in the US Patent and Trademark Office typically receives the patent. Thus, it is critical to establish a filing date in the US Patent and Trademark Office to protect your status as the first inventor to file a patent application on a particular invention.

The invention mining process is an important first step to be sure your valuable inventions don't slip through the cracks. Failure to protect critical inventions results in a loss of valuation and the inability to exclude others from using those inventions.

Brainstorming Sessions

Scheduling specific times for brainstorming activities is an important task for technology companies that rely on innovative ideas to grow and thrive. Individual and group brainstorming sessions can help uncover gold nuggets of innovation hidden throughout your team.

I've worked with many people who come up with valuable inventions during their brainstorming sessions. These people are often prolific inventors who schedule time every week to get away from their office and do some innovative thinking. These inventors typically keep a list of problems that need to be solved. They pick one of these problems for focus during a brainstorming session. This makes for a more valuable session because the person is focused on ideas related to an important problem that needs a solution.

Group brainstorming sessions are also valuable by allowing several team members to collaborate on solving a particular problem. An important rule for group brainstorming is: no criticism and no judging. When you are capturing creative ideas, do not criticize or judge any of the ideas. Any type of criticism or judging may discourage people from sharing their ideas, which stifles the creativity of the group. I've encountered countless situations where someone offered an initial idea that seemed completely unreasonable. But that initial idea caused another person to come up with a modification of the idea, which was further modified by someone else. After several modifications, the initial idea had evolved into a very practical and valuable idea that became a new product feature, some of which were patentable inventions.

In my experience working with technology companies, it is important to encourage team members to participate in both individual brainstorming sessions and group sessions. These are different activities that both produce valuable results. Individual brainstorming without interruptions can produce strong initial ideas. The collective value of a group can further refine individual ideas based on the different knowledge and experience of the group members. It's valuable to bring ideas created during an individual session to a group session for further ideation.

A critical task for all types of brainstorming activities is: record everything. Even the smartest inventors can forget important ideas. For in-person brainstorming sessions, a white board or flip chart are helpful to capture ideas during the session. For virtual sessions, ideas can be recorded into a document that's shown to all participants via screen sharing. Some companies record the sessions using audio or video for future reference. You can use any recording technique; just be sure that every idea is recorded so you can track the development of those ideas. At the end of a group brainstorming session, it is important to identify the most significant inventions for further activity and assign a team member to take responsibility for that invention.

Invention Teams

If you are just starting to build your portfolio of patents, attempting to launch a company-wide invention program may disrupt the day-to-day operation of your business. In many cases, your team is already overloaded, and a large invention program can be met with resistance.

In this situation, consider starting small with an initial invention team that doesn't upset the daily activities of your business. As you see success with the initial invention team, you can gradually increase the number of invention teams and expand the invention activities throughout the organization.

It's important to select a diverse group of people to take advantage of the different perspectives of team members performing different roles within the organization. For example, when working with a new client, a small invention team was created with six members from different departments within the company – engineering/product development, sales, marketing, customer service, and the executive leadership team. The group started by identifying several problems with the company's existing products as well as gaps in the marketplace.

Over the course of eight weeks, the small invention team developed solutions to several problems and brainstormed a new product opportunity to fill a large gap in their industry. Two of the solutions were successfully implemented by the company to improve the features and operation of the company's flagship product. The new product opportunity that fills an industry gap is still in development and is expected to significantly increase company revenue when the new product is released.

Overall, the results produced by the small invention team increased product sales and generated positive media coverage of the new product features. Since the first invention team was successful, the company is adding more teams to generate an increasing flow of new inventions.

You can use this same approach in your organization. Start slowly with a small invention team, and gradually expand to continue identifying and developing valuable inventions for your company.

Now that you're developing a pipeline of new inventions, the next chapter will describe how to evaluate those inventions using criteria specific to your company. This evaluation will prioritize the most important inventions that are the best candidates for patent protection.

CHAPTER 4
EVALUATE INVENTIONS USING CRITERIA SPECIFIC TO YOUR COMPANY

"Business has only two functions – marketing and innovation." Peter Drucker

The previous chapters discussed several techniques to create a steady flow of new inventions. Now that you are identifying and creating inventions on a regular basis, it's time to discuss how to evaluate those inventions and identify the strongest ones that provide the most value to your company. Obtaining patents for strong inventions can increase company valuation and provide a valuable competitive edge. This chapter will explain how to create a diverse invention evaluation committee, set criteria for evaluating inventions, and evaluate the inventions based on that criteria.

In my experience with hundreds of technology companies, I have seen many situations where companies had no system for evaluating inventions. I've even seen cases where a company selected an invention to patent because they wanted to "reward" an employee. Although rewarding an employee is a good thing, this is a terrible way to manage valuable company assets such as patents.

If you really want a strong patent portfolio, you must evaluate all inventions generated by your team. Don't ignore this important activity. The value of your patent portfolio depends on the quality of your evaluation. Otherwise, you are likely to invest significant time and money to prepare and process patent applications that provide minimal value to the company.

Create Your Invention Evaluation Committee

When evaluating inventions, get input from people throughout your organization, including people who perform different roles in the company. As I discussed in Chapter 3 regarding brainstorming and identifying inventions, people with different roles in your organization have unique perspectives regarding the potential value of the invention, the cost to develop and produce the invention, and whether the invention is consistent with the company's short-term and long-term goals.

Here are a few examples of people to include in your invention evaluation committee:

- *Company executives.* Company leaders can provide guidance on which inventions are aligned with the organization's goals and growth plans.
- *Engineering/product development team.* The technical team can determine which inventions are feasible and provide advice about the time and cost to develop and launch a particular invention.
- *Sales/marketing team.* These team members can offer input regarding which inventions are most likely to impact sales. For example, inventions that address a feature customers have been asking for are likely to have higher value to the company than inventions related to features that are not in demand.
- *Customer service.* Your customer service team can identify common problems and other customer feedback related to existing products. This information is important to consider when evaluating new inventions. If an invention solves a frequent complaint, it can be a valuable invention for the company.

By building a diverse invention evaluation committee, your company benefits from the unique experiences and varying perspectives of the committee members. This diversity combined with the evaluation criteria discussed below provides significant information to evaluate and prioritize inventions so the company is protecting the most valuable inventions.

Define Your Invention Evaluation Criteria

To effectively evaluate your team's inventions, you want to provide specific criteria that is applied by the invention evaluation committee. These criteria are specifically selected based on your particular organization, its goals, growth plans, and other factors discussed below. Here's a list of evaluation criteria to consider for your own invention evaluation committee.

- *Feasibility of the invention.* Analyze the invention to determine whether it is commercially feasible to design and produce the invention. For example, an invention may not be the right choice if the cost is too high, the development time is too long, or current technology is not sufficiently advanced to design a useful product.
- *Must-have feature.* Determine whether the invention is related to a must-have feature that customers are

asking for or that fills a gap in the marketplace. If so, this may be a valuable invention if the other criteria are confirmed.

- *Company goals.* Identify the company's short-term goals and long-term goals. Specifically, consider goals related to company growth, new product development, and desired investments from angel and venture capital investors. Consider how each invention satisfies the goals of the organization.
- *Exit strategy.* Consider the company's exit strategy and how inventions can support that strategy. For example, if the exit strategy is to seek an acquisition of the company, building a portfolio of patent assets may increase the valuation of the company at the time of acquisition. In this example, when evaluating inventions, consider the likelihood that a patent covering the invention would increase the company's valuation.
- *Potential patentability problems.* Some inventions may be valuable, but difficult to protect with a patent. For example, some software-based systems are difficult to patent in the United States and other countries. For a high-value invention, these potential problems may be worth the risk. However, for less valuable inventions, it may not be worth the

time and money to pursue the invention in light of the potential patentability problems.

- *Easy to design around.* Determine whether the invention is easy to modify and avoid infringing a patent on the invention. For example, if the invention is a specific implementation and there is at least one way to effectively design around the invention, any patent covering the invention would likely have little value. Preferably, you want to patent inventions that provide broad coverage and are difficult to design around.
- *Easy to determine infringers.* If you obtain a patent on the invention, can you easily determine if a competing product infringes the patent? For example, if the invention is embedded deep in the software of a complex system, it may not be apparent whether a competitor is using the invention. In this case, a patent may have minimal value because you cannot identify infringers.
- *Value of invention to company.* In view of the company's goals and other criteria, what is the value of the invention to the company? Determine whether the invention provides a competitive advantage or protects against someone copying the invention. Evaluate both the short-term value and

the long-term value to the company. You may also consider whether the invention has potential licensing value that could create additional revenue for the company.

- *Defensive value.* Some inventions are patented for defensive reasons, as further discussed in Chapter 5. Defensive patents provide protection against overly aggressive competitors by reducing the likelihood of litigation and providing a cross-licensing opportunity if litigation occurs.
- *Should the invention be maintained as a trade secret instead.* For some inventions, it may be desirable to maintain the invention as a trade secret instead of disclosing the details of the invention in a patent application. For example, maintaining the invention as a trade secret may be desirable if the invention can be protected from public disclosure and it may be difficult to obtain a patent on the invention.
- *Lifetime value of patent.* Consider the lifetime value of the invention. Is the invention specific to today's technology and likely to be obsolete within a few years as technology advances? Since it often takes 2-3 years for a patent to issue, inventions with a short lifetime may not obtain patent protection during the lifetime. Alternatively, if the invention has broader

coverage and will remain valuable after technology advances, it may be worth patenting the invention (depending on other criteria).
- *Confirm ownership of the invention.* Determine whether there are potential ownership issues with the invention. For example, evaluate whether there's an inventor who is not associated with your company or some other issue that may prevent your company from being the sole owner of the patent. As with the potential patentability problems discussed above, you might still move forward if the invention has a high potential value.

Your company doesn't need to use all of the above criteria when evaluating every invention. Instead, use the criteria that is most relevant to a particular invention and appropriate for your organization.

All criteria are not necessarily weighted equally. For example, the feasibility of an invention may carry a high weight such that an invention that is not feasible is rejected regardless of whether other criteria are satisfied.

Implementing Your Evaluation System

After you have created your invention evaluation committee and defined your invention evaluation criteria, it's time to start reviewing inventions in your pipeline.

I recommend starting each evaluation meeting by reviewing the criteria that is important to your company. When I work with clients, we typically develop a checklist that identifies all of the company's evaluation criteria. The invention evaluation committee completes the checklist for every invention being evaluated.

It's important to consider all evaluation criteria for each invention. Even if an invention looks valuable after reviewing a few of the criteria, a single negative criterion may cause the invention to be rejected. For example, if an invention is feasible and fills a need in the marketplace, it may be rejected if it has potential patentability problems or is easy to design around.

After evaluating an invention, determine the next step. In some situations, the decision is final (accepted or rejected), and other times additional information is needed to decide. Here are some example actions to take in response to an invention evaluation:

- Do not pursue invention.
- Prepare a patent application to protect the invention.
- Defer the evaluation decision to a future date.
- Request additional information from the inventors.
- Request additional analysis by other team members.

- Schedule a brainstorming session to discuss invention further and possibly broaden the scope of the invention.

Visit SteveSponseller.com/tools to get an Invention Evaluation Checklist template you can use in your own company.

Be sure to record the next action for each invention evaluation, and identify a lead person responsible for overseeing that action. Record these actions and the lead person in the invention tracking system mentioned in Chapter 2.

After each meeting of the invention evaluation committee, continue collecting new invention disclosure forms in preparation for the next committee meeting. Future meetings will review open action items from previous meetings and discuss new inventions received since the last meeting.

You now have an invention evaluation committee that reviews new inventions using a set of criteria that is specific to your company. In the next chapter, I will help you start building a portfolio of patents that protect the four critical categories of inventions.

CHAPTER 5
Protect the Right Inventions

"Necessity is the cause of many inventions but the best ones are born of desire." Guglielmo Marconi

The last chapter explained how to create your invention evaluation committee, define your evaluation criteria, and launch your evaluation system. After your invention evaluation committee is up and running, it's time to learn how to protect the right inventions. A patent portfolio full of mediocre patents offers minimal value. But a strong patent portfolio filled with patents covering the four critical categories of inventions provides significant value and protection to your company. This chapter will help you protect the right inventions that create a broad and valuable patent portfolio.

All Patents Are Not Created Equal

Some patents are extremely specific and provide protection for a particular design implementation. Unfortunately, many of these specific patents have minimal value if someone can easily design around the patent. Other patents provide broader protection and cover multiple design implementations. These broader patents are typically more valuable because they provide a larger scope of protection against competitors.

The evaluation process discussed in Chapter 4 helps you identify high-value inventions and weed out the weaker inventions, such as ones that are easy to design around or have a short lifetime. The evaluation criteria are designed to identify inventions that support broad patent protection. Be sure to use that criteria when evaluating inventions related to the four categories discussed below. It will save you time and money.

Four Critical Categories of Patents

As you build your patent portfolio, it's important to include inventions that cover the four key areas discussed below. Patents covering these inventions will collectively provide a barrier to entry for competitors and position your company as an innovator in its industry. The four critical categories of patents discussed here individually are: core technology, critical must-have features, future inventions, and defensive patents.

Core Technology

Identify the critical inventions in your core technology. This could be your novel computing framework, an artificial intelligence platform, or any other unique system that is the foundation of one or more products. Often, multiple products (and multiple inventions) are built on this core technology. Protect inventions related to your core technology as quickly as possible to safeguard those important systems.

For example, suppose your company has developed a new health monitoring system that collects data from multiple sources and performs a unique analysis of the collected data. The manner in which the data is analyzed is novel and sets the company apart from others in the industry. In this example, the data analysis system is the core technology for the company. Since this analysis is the company's "secret sauce," it's a strong candidate for patent protection.

For many small companies, it's important to focus on protecting their core technology before moving on to the other categories discussed below. This core technology distinguishes the startup from existing competitors in the marketplace. Patenting those inventions can provide a competitive advantage by preventing competitors from copying the core technology.

Timing is critical when protecting your core technology and other inventions. Unfortunately, I've met with many technology leaders who didn't act fast enough to protect their core system. In some cases, they had sold their product or publicly disclosed details of the core technology more than a year before our first meeting. In the United States, a sale of the product (or a public disclosure of the invention) starts a one-year deadline to file a patent application. If you don't file your patent application within one year of the sale or public disclosure, you lose your opportunity to file for patent protection on that invention.

Don't make this mistake! Keep this patent deadline in mind when launching a new product or making other public disclosures.

See Chapter 8 for more examples of key events that require patent action.

Critical Must-Have Features

In many situations, one or two product features are responsible for making a product stand out in the marketplace. Even if these features are not part of the core technology, protecting these features can provide a strong competitive advantage. This is a case where information from the sales and marketing teams is valuable to identify product features that your customers love.

For example, if your home automation platform has a unique feature that allows other devices, such as smart appliances and other IoT (Internet of Things) devices, to interface with the platform quickly and easily, that feature may be valuable. If you can patent the system that supports the quick and easy interface with the home automation platform, you can exclude competitors from using the inventions protected by the patent. If this is a must-have feature for customers, an issued patent gives you a strong advantage in the market.

Schedule an activity with your team to identify all of the must-have product features in your existing products as well as all products in development. Get input from the sales team and marketing group to learn which features are most popular and which features have been on customer wish lists.

Future Inventions

Although it is important to protect your core technology and key features, don't limit yourself to protecting today's technology in your current products. You can significantly increase the value of your patent portfolio by identifying and patenting inventions that will likely be essential in the future.

Look at today's trends, such as trends in your own industry as well as worldwide trends affecting many areas. Evaluate these trends and look at ways to create new products or modify existing products to gain alignment with these trends.

For example, if a current trend in your industry is an increased use of wearable devices, such as smart watches and fitness trackers, consider how your company's products can better engage with customers in an intimate manner through a wearable device. Look for ways to provide an improved customer experience through valuable features and other inventions.

Here's one approach to identify future inventions. Evaluate various trends that may impact your company and your industry. Then, take another step by looking for new problems that will be caused by these trends. The problems may not exist yet, but your creative team can look into the future and predict these problems ahead of time. By anticipating these upcoming problems, your company can begin creating solutions to problems that don't exist yet. But, when the problems do arise, your company can be positioned to remedy those problems with a patented solution. Since you solved the problem before it even existed, you have a good opportunity to get broad patent protection for your solution, which provides a strong competitive edge.

Here's an example scenario. As wearable devices become more powerful, they may control and interact with a larger variety of systems. For example, a future wearable device may control health-related systems such as pacemakers and insulin pumps. It will be important for these future wearable devices to be sure they are controlling the pacemakers and insulin pumps accurately. This means the wearable device must be able to identify the person wearing the device with 100% accuracy. If the wearable device is worn by a different person, the wearable device must realize that situation and avoid accidentally controlling any health-related systems associated with a different user. Thus, there are many potential problems with wearable devices as they become more powerful. Those future problems can be predicted today, and the company's team can begin solving those problems right now so that practical solutions are already available (and covered by a patent) when the problems arise.

Defensive Patents

When building your patent portfolio, it's important to include some patents that provide defensive protection against aggressive competitors. For example, a large competitor may sue a smaller startup for patent infringement in an effort to eliminate the startup as a competitor. When selecting a litigation target to sue, these large competitors typically evaluate the startup to see if they have any patents that could be asserted against the large competitor. If the competitor is considering multiple startups as targets, the startup with no patents (or nothing that applies to the competitor) is more likely to be selected as the litigation target.

By including patents in your portfolio that cover the competitor's products or services, you have created an opportunity to settle any litigation by cross-licensing patents. This type of cross-licensing allows each entity to use the other's patented technology based on the terms of the agreement. This can be a mutually beneficial solution to the litigation. However, if your company has no patents that can be asserted against the competitor, then you aren't bringing anything to the bargaining table. Thus, part of your patent strategy should include patenting inventions that can be asserted against your key competitors' products.

Some companies only focus on patenting inventions that protect their own products or services. This is a mistake. Although those patents are important, you should set aside some time to patent inventions that cover your competitors' systems. There's no patent office rule or other requirement in the United States that you can only protect inventions related to your own products and services. Also, there's no rule in the United States that you must produce a product that contains the patented invention. You are free to obtain patent protection for any invention, including those that you do not intend to develop into a product offering.

These defensive patents are an important part of any valuable patent portfolio.

You have worked hard to assemble a team of talented people who create innovative products and give your company an advantage in the marketplace. Protect that investment in your team by protecting your core technology, critical must-have features, future inventions, and defensive patents.

After reading this chapter, you should now have a better understanding of how to identify and protect the right inventions that provide strong value for your company. The next chapter will discuss additional techniques for building a portfolio of patents that's filled with high-value patent assets.

STEVEN SPONSELLER

CHAPTER 6
BUILD A PATENT PORTFOLIO FILLED WITH HIGH-VALUE PATENTS

"I believe innovation is the most powerful force for change in the world." Bill Gates

Now that you understand the four key patent categories (core technology, critical must-have features, future inventions, and defensive patents), this chapter will discuss the mechanics of building a strong patent portfolio that contains a growing number of valuable patents.

As mentioned earlier, a patent portfolio full of mediocre patents offers minimal value. But hand-picking specific patents that fall within one of the four key patent categories will produce a strong and secure patent portfolio. This chapter shows you how to do that.

The Patent Application Process

The patent application process begins by preparing a patent application for an identified invention, such as an invention approved by the invention evaluation committee. Before drafting a patent application, a patent attorney (or patent agent) typically meets with the inventors and reviews documentation related to the invention. After obtaining the necessary invention details, the patent attorney prepares a first draft of the patent application. The patent application is reviewed by the inventors, and revisions are made until all inventors are satisfied with the patent application.

A final draft of the patent application that has been approved by all inventors is then filed with the US Patent and Trademark Office. The patent application is assigned to a patent examiner and placed in the examiner's queue. Eventually, the patent examiner reviews the patent application and searches for public documents that describe similar inventions. The patent examiner issues a written Office Action that accepts or rejects the application. If the application is initially rejected by the patent examiner, that's the start of a negotiation process. The patent attorney can make certain revisions to the patent application and argue that the patent examiner's rejection is incorrect. This negotiation process continues until the patent examiner approves the patent application or the inventors decide to let the patent application go abandoned.

If the patent examiner approves the patent application, it will be assigned a patent number and officially issued by the US Patent and Trademark Office. Once the patent has issued, it becomes another asset in the company's growing patent portfolio.

How to Use Provisional Patent Applications

A "traditional" patent application has specific rules regarding a written description, formal drawings, and claims that define the legal protection provided by a patent that issues from the application. This is the type of application most people are referring to when they mention "patent applications" – and it's the type of application referred to in the preceding section when discussing the patent application process.

However, there's another type of application referred to as a provisional patent application. The provisional patent application allows the inventors to establish a filing date for their invention in a fast and cost-effective manner. The provisional patent application is not examined and never becomes a patent itself. However, it is a powerful tool for quickly securing a filing date that gives the inventors time to decide whether to file a more expensive (and more complex) traditional patent application.

Establishing a filing date for your invention is important because the United States is a "first inventor to file" country for patent applications. So, if two people create the same invention at approximately the same time, the first person to file a patent application (including a provisional patent application) with the US Patent and Trademark Office is given the opportunity to protect the invention with a patent. In this "first inventor to file" system, it doesn't matter who creates the invention first; the winner is the person who gets the earlier filing date.

Thus, provisional patent applications are useful in establishing a filing date in a fast and cost-effective manner. Provisional patent applications also give the inventors time to determine the commercial viability of an invention. After filing a provisional patent application, the inventors have 12 months to decide whether to file a traditional patent application. This 12-month time period lets the inventors continue developing the invention, seeking investors to support the invention or company, and testing the invention in the marketplace. If the invention proves valuable, the inventors can file a traditional patent application within 12 months of filing the provisional patent application. In this situation, the traditional patent application will receive the earlier filing date assigned to the provisional patent application.

Additionally, once you have filed a provisional patent application, you can use the "Patent Pending" designation on your product. You can continue to use this "Patent Pending" designation for 12 months after filing the provisional patent application and continue after those 12 months if you file a traditional patent application.

Provisional patent applications have simplified rules that streamline the drafting and application process, thus saving time and money. When establishing a filing date is critical and funds are tight, a provisional patent application is a good option to consider. Chapter 8 discusses examples of specific situations where a provisional patent application is particularly useful.

Cover All Four Critical Categories

As discussed in Chapter 5, it's important to protect your core technology, critical features, future inventions, and defensive patents. During your patent evaluation process, look for inventions that fit into each of these four categories, and protect those inventions to create a strong and diverse portfolio of patents.

If you find that your team is not generating inventions in one or more of the four categories, take steps to increase the inventions in those areas. Consider brainstorming sessions, creating invention teams, and conducting invention mining, as discussed in Chapter 3 to increase the submission of new inventions in those categories.

Additionally, Chapter 7 discusses several ways to boost the identification of new inventions that lead to broader patent coverage.

Keep the Invention Pipeline Full

To continually add new high-value patents to your patent portfolio, you need a steady stream of new inventions in your invention pipeline. Without a full invention pipeline, the invention evaluation committee doesn't have a good selection of inventions to choose from. If there aren't enough high-quality inventions to evaluate, the committee may not select any to pursue, which stalls the growth of your patent portfolio.

You can keep your invention pipeline full by using the techniques discussed in Chapter 3, and begin incorporating an invention mindset throughout your organization. This invention mindset (or invention culture) is important for technology companies that rely on new inventions to grow and thrive. Encourage team members to practice individual and group brainstorming sessions to generate inventions as part of their day-to-day activities. Consider an invention contest or other activity discussed in the next chapter to help stimulate creation of more inventions.

Visit SteveSponseller.com/tools to get a One-Page Infographic summarizing the three-step system described in this book.

By applying the topics discussed in this chapter, you are building the foundation of your patent portfolio. In the next chapter, you will learn additional "patent booster" techniques to accelerate the growth and value of your patent portfolio.

STEVEN SPONSELLER

CHAPTER 7
Patent Boosters

"Logic will get you from A to Z. Imagination will get you everywhere." Albert Einstein

In the preceding chapters, I discussed a system to start building a patent portfolio filled with high-value assets. Now I will provide some techniques that will help expand the content and value of your patent portfolio.

Create an Invention Culture

As mentioned in Chapter 6, an invention culture is important to maintain a steady flow of new inventions for the invention evaluation committee to review. As you build an invention culture, your team members will incorporate innovative thinking into their daily activities. An invention assessment can become a required step for all projects, where the team specifically discusses inventions developed during the course of a project. You can suggest a "question of the week" and ask all team members to provide innovative answers to the question.

For example, one of my clients recently wrote the following "question of the week" on a white board in the cafeteria: How can we modify our user interface so a brand-new user can start using our product within five minutes? The company requested input from the entire team across all departments. The results obtained from this question included several valuable ideas that simplified the user interface and made it more intuitive to new users. These ideas were not patentable but significantly improved the company's product and made the customers happy. This exercise also gets employees thinking of new inventions.

I've worked with many companies that incorporate inventions into each team member's annual evaluation. This emphasizes the importance of inventions to the team's daily activities. These types of programs must be supported by the company leaders to underscore the value of innovative thinking throughout the company.

Invention Contests

Although brainstorming sessions can generate great ideas, they tend to get stale if you repeat the same format every time. If you want to foster teamwork and develop an invention culture that produces a flood of innovative ideas, you need to spice up your repetitive brainstorming events.

Here's a strategy I've used with technology companies to gamify the invention process. Create an invention contest that includes various creative activities that make idea generation fun. This is a great diversion from day-to-day work activities and gives your team a chance to exercise the creative part of their brains.

Try this example approach to implement an invention contest in your organization.

Invention Teams

Start by creating multiple invention teams with 4-8 people on each team. If possible, each team should have a cross-section of the company's departments. For example, each invention team may have at least one person from sales, marketing, customer service, and engineering/development. This diversity of company roles will produce a broader range of ideas by receiving perspectives from different parts of the company.

The first task for each invention team is to create a clever name for the team. This is the first step in the invention contest because points are awarded to the team with the best name.

Define the Invention Contest

To kick off the invention contest, explain the purpose of the contest, the contest rules, and how teams can earn points during the contest. And, most importantly, explain the prizes for the top invention teams. I recommend selecting significant prizes – things that are highly valued within your organization. One company I worked with gave the winning team members custom jackets embroidered with "Invention Champion" and an attractive logo. There were lots of bragging rights for the owners of these jackets! If you want ideas for valuable prizes that will motivate the teams, just ask your team members for ideas.

Your company can host a one-day invention contest or spread the contest across several weeks, with multiple activities during the contest. Several weeks allow more time for the invention teams to work on their ideas and develop stronger inventions.

Contest points are awarded in many ways. Points may be awarded for each invention identified by the team. Judges can evaluate the inventions and award points based on the creativity of the invention and its potential value to the company. You can also award points for the funniest invention, most out-of-the-box idea, and other categories.

Contest Activities

As the contest begins, each invention team identifies problems that need to be solved. In some contests, the company leaders may provide a few problems that they want the teams to focus on solving. These problems could be issues with the company's existing products, ideas to modify existing products to fill a particular gap in the marketplace, or ideas for new products that address specific trends in the industry.

Typically, each invention team selects a particular problem to solve during the contest. The goal of the contest may be to identify as many solutions as possible, with the team getting points for each solution. In other contests, the goal is to identify up to three solutions that the team believes are the best. The "best" invention can be defined by various criteria, such as some of the invention evaluation criteria discussed in Chapter 4.

In some contests, teams are required to use different techniques for developing initial solutions to a problem, then refine those initial solutions through additional brainstorming and ideation.

Scoring

As mentioned above, contest points may be awarded in many ways. In addition to earning points for the number of solutions or the best invention, companies can award points based on any other categories. Some companies have a panel of judges who award points. Other companies have all contest participants vote on the best inventions (but team members cannot vote for their own team's ideas).

Prizes and Celebration

Select prizes that are valuable to people in your organization. A well-organized invention contest can generate multiple high-value inventions. Offer prizes that recognize the value of the winning team's invention. These prizes help encourage participation during the current contest and entice employees to actively participate in future contests. Get input from your team members before the contest to select motivating prizes.

Celebrate everyone's efforts and thank them for participating. Some companies host contest luncheons after the event has finished and give gift cards to local restaurants and movie theaters for all team members who participated in the contest. Celebrate your team's efforts in a manner that's most appropriate for your organization.

Ongoing Results

The goal for an invention contest is to get teams of people from different parts of the organization working together to solve problems in a fun environment. This can improve morale and create a strong team environment. The activities give your team a fun break from their usual work routine. Invention contests and other events often generate many new ideas that can lead to new products and improvements to existing products.

In my experience, the creation of innovative ideas doesn't end when the contest is over. The contest produces a creative spark in the contest participants so they continue thinking of new ideas over the coming days and weeks. I frequently hear stories from team members who "kept thinking about one of the problems from the invention contest and just came up with a fantastic solution that . . ."

Consider making invention contests a regular activity in your business to enjoy a boost in valuable new ideas and a stronger innovation culture. It's also a fun team-building exercise. In the next chapter, I will discuss several critical events that require patent action. These events can cause a loss of patent rights if not handled properly.

STEVEN SPONSELLER

CHAPTER 8
CRITICAL EVENTS THAT REQUIRE PATENT ACTION

"Innovation distinguishes between a leader and a follower."
Steve Jobs

The preceding chapters discuss a system to build a strategic patent portfolio that contains highly valuable patents. After working with hundreds of technology clients, I've identified a list of critical events that require companies to take action to preserve their patent rights. If you fail to take appropriate action, some of these events can prevent you from protecting your inventions in the United States and/or foreign countries.

The last thing you want to hear is, "I'm sorry, but it's too late to patent that invention."

I've had to share this bad news with far too many business leaders during the past 20 years. This is terrible news for your company, especially when a critical invention is involved.

Even worse, most of the time the problem could have easily been avoided.

The 12-Month Deadline

The United States patent laws have specific timing requirements for filing a patent application. If a company publicly discloses an invention or sells a product containing the invention, those activities may trigger a 12-month patent application filing deadline.

In most situations, if a patent application is not filed by the 12-month deadline, the company forfeits the ability to protect the invention with a patent.

Unfortunately, many tech company leaders don't understand these patent filing requirements. They recognize the importance of patenting inventions but often let the filing deadlines slip through the cracks because they aren't aware of the 12-month deadline.

That's when I become the bearer of bad news.

Sometimes, though, companies get lucky.

In one case, a company's product was first sold almost a year before our initial meeting, but we still had a little time to get a patent application filed before the 12-month deadline.

This was a great outcome. But if our initial meeting had been scheduled just one week later, there would be no patent for that valuable invention. It turns out the invention is critical to the company because it's part of their core technology that is disrupting the data communication industry.

Invention Triage™ Process

Fortunately, there's a solution to forfeiting your patent protection – and it doesn't even require being lucky!

The solution is my Invention Triage™ process.

How does it work? By identifying inventions and determining which ones are the most urgent and need to be handled first. For example, inventions that are approaching a 12-month deadline are the highest priority.

The Invention Triage process is particularly important when you are just starting to build a patent portfolio, or it has been several months since your team identified or evaluated recent inventions.

I recently started working with a computer hardware company that already had four issued patents covering inventions developed several years ago. However, after filing the initial patent applications, the company stopped focusing on protecting new inventions.

Even though several new products were developed and released during the previous two years, no thought was given to identifying and protecting inventions in those products.

Our first activity was to apply the Invention Triage process to identify and prioritize all inventions developed since filing the initial patent applications. This process yielded five important inventions that distinguish the company's products in the marketplace. We analyzed all of the inventions and had to eliminate three from consideration because they were past the 12-month patent filing deadline.

The remaining two inventions were carefully evaluated using criteria specific to the company's goals and priorities. The evaluation determined that both inventions represented significant value for the company, so they decided to file patent applications for both inventions to strengthen their portfolio of patents.

Although three of the inventions were already past the 12-month deadline, the company was able to rescue two of the inventions. If they had waited two more months to implement the Invention Triage process, all five of the inventions would have been forfeited due to the 12-month patent filing deadline.

An effective Invention Triage process includes analyzing all inventions that have been:

- released in a product within the last 12 months
- offered for sale within the last 12 months

- described in any public disclosure (such as white papers, public demonstrations, or press releases) within the last 12 months

An important part of the Invention Triage process is scheduling meetings with your development team to identify any inventions that may be approaching the one-year patent filing deadline. If you identify any products or product announcements in any of the categories above, it's important to fully evaluate these inventions to determine whether to file a patent application quickly. Use the same evaluation criteria and process discussed in Chapters 4 and 5.

Don't postpone this important activity. You may have at-risk inventions that need immediate attention.

Repeat the Invention Triage process each time there has been a gap of more than a few months in your invention identification and evaluation activities.

Protect Your Patent Rights *Before* Public Disclosures

The following list is related to activities that may accidentally (or sometimes intentionally) disclose an unprotected invention. If you have not filed a patent application before performing these activities, you may give up your right to file a patent application in foreign countries. Other activities can eliminate your ability to file a patent application in the United States.

As mentioned above, certain types of activities won't prevent you from filing a patent application in the United States but trigger a one-year deadline within the United States for filing a patent application. For these activities, you still have the right to file a US patent application, but you must file the application within one year of the triggering event.

Here are five activities you want to be aware of to avoid accidental loss of patent rights:

1. *Trade shows.* Demonstrating products or services that contain an invention can cause a loss of the ability to file a US and/or foreign patent application. Before attending any trade show (or conference), determine whether your company plans to present or demonstrate inventions that are not yet protected. If unprotected inventions may be discussed or demonstrated, your company should consider taking steps to file a patent application before the trade show to preserve both US and foreign patent rights.

2. *White paper publication.* Before publishing a white paper, article, or other public document, check to see if any inventions are disclosed in the document. If so, consider

filing a patent application before publishing the document to protect the identified inventions.

3. *New product/service announcements.* Any type of public announcement, such as press releases and broadcast interviews, should be evaluated to determine whether any unprotected inventions will be disclosed. As noted above, you should consider protecting any inventions being discussed prior to the public announcement.

4. *Advertising campaigns.* Similar to the public announcements discussed above, evaluate your advertising campaigns to be sure they don't disclose an invention that is not protected. If an invention is disclosed in the advertising campaign, consider changing the advertisement to avoid mentioning the invention or filing a patent application before running the advertisement.

5. *Public demonstrations or sales presentations.* If you are planning any type of public demonstration or sales presentation, determine whether the demonstration or presentation will disclose any important inventions. Also, evaluate whether the demonstration or presentation is an offer to sell a product or service that includes an invention. In either of these situations, you want to consider protecting your invention prior to the demonstration or presentation. As mentioned above, if you offer to sell a product containing an

invention, you still have the right to file a US patent application, but you must file the application within one year of the offer to sell.

I have seen too many companies accidentally give away valuable inventions because they publicly disclosed the inventions before they were properly protected. In other cases, companies offered to sell a product containing an invention and failed to file a patent application within 12 months, causing them to forfeit the ability to protect the invention. You can avoid these problems by identifying inventions that will be disclosed during any of the five activities mentioned above. Then, decide whether any steps are needed to protect those ideas prior to public disclosure.

Let everyone in your organization know about the potential problems if any of these five activities occur without first protecting any important inventions.

If you want to quickly file a patent application to protect a key invention before a public announcement or demonstration, a provisional patent application (discussed in Chapter 6) is a good tool for this situation.

Inflection Points

The previous section discussed protecting your patent rights before making a public disclosure of an invention. Now, let's discuss several inflection points in your business that demand a patent analysis.

An inflection point is a time when a significant change occurs (or is likely to occur). Every business reaches inflection points where change is inevitable. Failure to act at these points can result in a loss of momentum or failure of your business. Although there's no wrong time to evaluate your inventions, there are several critical inflection points that demand a thorough patent analysis.

Here are five inflection points that should trigger an analysis of your current patent portfolio:

1. *Expanding into a new market.* When your business moves into a new market, it is competing with other companies that have an established customer base. To gain traction in the new market, you want to distinguish your company from the existing competitors. Identify the features of your product that are innovative and stand out from the competitors' products. Before entering the new market, be sure you have protected any inventions that will give you a competitive advantage in that new environment.

2. *New product launch.* New products give your company an opportunity to increase revenue and reach more customers. When developing a new product, don't just bring

in your existing development team. Involve people throughout the organization in the invention process, such as employees in the sales, marketing, and customer service departments. Including these members of your team provides a broader view of the industry and customer needs, which can generate more innovative and more valuable inventions. Use the new product launch as an opportunity to further distinguish your company and its products from competitors. Make your company stand out as an industry innovator. In my work as an intellectual property attorney, I have seen many situations where a single new product is responsible for the exponential growth of the company. So, don't just create another "me too" product; launch something that will generate a powerful buzz in the industry. And be sure to protect the inventions in that product to further strengthen your patent portfolio.

3. *Changes in your industry (industry trends).* Changes in your industry and overall market trends may require you to adapt your products to meet these changes and adjust your strategy to leverage the new market trends. For rapid industry changes, you want to swiftly develop innovative ideas to modify existing products based on the changes. For slower changes, such as evolving market trends, focus your invention activities on the new problems that will be created by the market trends. Develop inventions that solve those problems

so your company already has solutions to the problems when they arise. These solutions put you a step ahead of the competition because you anticipated the problems and solved them in advance. Protect the inventions in your solutions to further strengthen the "future inventions" category of your patent portfolio, as discussed in Chapter 5.

4. *New competitors entering your market.* A new kid on the block is the perfect catalyst to develop features that can set your product apart from the new competition. If your business has been operating for a while, you already have an existing customer base. Leverage that advantage by obtaining feedback from your current customers to identify ways to improve your products. Examine common problems handled by your customer service department; these problems represent opportunities to invent. Use your existing position in the market to your advantage, and find ways to distinguish your product offerings from the new competitor. It's important to act quickly to engage your innovation team and identify ways to preserve (and hopefully expand) your market share. The new inventions you develop will enhance the value of your existing patent portfolio.

5. *Seeking investment capital.* Many organizations seek to raise capital at various points during the growth of the company. Savvy investors evaluate a company's competitive

position in their market and the protection of proprietary systems and procedures. If your company is anticipating a need for a capital investment, consider boosting your innovation efforts to develop inventions that make your products stand out from the competition. As discussed in earlier chapters, these inventions may include new products or improvements to existing products by adding new must-have features. Filing patent applications to protect these inventions provides a stronger portfolio of patents that can improve your company valuation and the attractiveness of your company to investors.

If your business is approaching any of these inflection points, start planning your innovation activities today so you can successfully navigate through the important transitions.

As discussed in earlier chapters, your business needs a system for regularly identifying, evaluating, and protecting inventions. But there are times when your business needs an extra boost of innovation to break through an inflection point. Recognizing these inflection points and responding with innovative thinking will accelerate your business growth to a new level. Watch for these five inflection points, and take action when they occur.

The critical events discussed in this chapter are important triggers for all technology companies (as well as companies in non-tech industries). I recommend that you work with your team to understand the importance of these events and put in place systems to avoid accidentally disclosing unprotected inventions and properly handle inflection points.

During my 20+ years of experience as an intellectual property attorney, I've seen many patent-related mistakes occur repeatedly. In the next chapter, I will discuss the seven most common mistakes I see and explain how to avoid those mistakes in your own company.

STEVEN SPONSELLER

CHAPTER 9
PATENT PITFALLS TO AVOID

"Failure is the opportunity to begin again more intelligently."
Henry Ford

This book has described a three-step process to build a valuable portfolio of patents. The process includes (1) identifying your inventions, (2) evaluating all inventions, and (3) protecting the right inventions. Although this three-step process is relatively simple, there are several important activities in each step. And your company needs to understand the rules discussed in the book to be sure your investment of time and money is actually building a portfolio of high-value patents.

After working with hundreds of companies and more than 1000 inventors, I have seen too many mistakes related to identifying and protecting important inventions. I've listed the seven most common patent pitfalls below and provide guidance so your company doesn't make any of these mistakes.

1. No System To Identify and Track Every Invention

Successful businesses have a system for identifying and tracking inventions. Since many inventions solve problems or fill market gaps, the invention process starts by finding opportunities to develop innovative solutions. Most inventions are incremental improvements that build on what's been done before, not revolutionary new discoveries. Small changes that solve real problems can produce significant value for your company.

Encourage your team members to identify problems, unmet needs, and market gaps. Then, let your team create inventions that address those issues. Schedule brainstorming sessions and other activities to ignite the collective creativity of your entire team. Be sure to implement a system to keep track of all those great ideas. Use a simple invention disclosure form and a tracking tool to be sure you capture all of the innovative ideas developed throughout your organization.

Review Chapters 2 and 3 for more information on identifying and tracking inventions.

2. Random Acts of Patenting (Without Proper Analysis)

Identifying new inventions is an important activity for all technology companies. But you must evaluate all of those inventions and take action to protect only the most valuable inventions. This is necessary to create a competitive advantage and prevent your critical inventions from being stolen.

Don't fall into the trap of randomly filing patent applications without any analysis of the invention's value to the company. This typically results in time and money being wasted to obtain a mediocre patent. Instead, create an invention evaluation group to review invention disclosures and evaluate each invention based on a specific set of criteria customized to your business goals. This approach will create a portfolio of patents that protects your company and provides a strong competitive advantage.

Review Chapters 4 and 5 for more information on evaluating inventions and selecting the best inventions for filing patent applications.

3. Not Leveraging Provisional Patent Applications

Establishing an early filing date for your invention is important in the United States, which is based on a "first inventor to file" patent system. The "first inventor to file" system means that if two inventors develop the same invention, the first inventor to file their patent application in the US Patent and Trademark Office typically receives the patent. Thus, it is critical to establish a filing date in the US Patent and Trademark Office to protect your status as the first inventor to file a patent application on a particular invention.

Provisional patent applications provide fast and cost-effective protection for important inventions. After filing a provisional patent application, your company has 12 months to decide whether to file a traditional patent application. This 12-month time period lets you continue developing the invention, seeking investors to support the company, and testing the invention in the marketplace. If the invention proves valuable, your company can file a traditional patent application within 12 months of filing the provisional patent application. In this situation, the traditional patent application will typically receive the earlier filing date assigned to the provisional patent application.

Review Chapter 6 for more information on using provisional patent applications in your business.

4. Failing to Celebrate and Reward Your Inventive Team Members

Innovation does not have to be boring! Creativity and innovation can be fun and should be rewarded. Schedule fun events like invention contests, themed brainstorming sessions, and other activities that give your team a break from the day-to-day repetitive tasks. Reward innovative thinking, and celebrate exciting new inventions.

Consider offering prizes and other rewards for your most innovative team members. Select prizes that are valuable to people in your organization. The company can receive significant value from the inventions, so make the prizes something of value that will encourage strong participation in the innovation process. If you want ideas for valuable prizes, ask your team. Find out what your team members consider most valuable, and provide those items as prizes (within reason).

Review Chapter 7 for more information on encouraging and celebrating invention activities.

5. No Agreements to Ensure Intellectual Property Ownership by the Company

Without proper agreements, your company may not be the exclusive owner of its intellectual property. To avoid invention ownership problems, use signed agreements that require all employees, contractors, consultants, freelancers, and others to assign their intellectual property rights to the company.

It's important to avoid the situation where an inventor is not required to assign their rights in an invention to your company. If this occurs, the non-assigning inventor may use the invention independently of your company and might offer the invention to your competitors without needing your approval. This would severely undermine the value of the invention to your company.

Establish company policies that require all employees, contractors, consultants, freelancers, and others to sign agreements that assign their intellectual property rights to the company.

6. Not Scheduling Fun Brainstorming Activities and Invention Contests

I admit that I'm guilty of hosting boring brainstorming sessions in the past. But I've learned how to make brainstorming fun so that people want to attend and participate. And invention contests are a fun alternative to brainstorming sessions.

Instead of scheduling a typical brainstorming luncheon with free pizza, try something different. A different location, different themes, or gamify the activity. Getting inventors out of their usual location can help stimulate creative thinking. Adding some games or creative activities also helps encourage innovative ideas. For example, have people warm up with a creative toy, a Rubik's Cube® puzzle, or a pile of Lego® blocks.

An invention contest is a fun way to foster teamwork and generate new inventions. The invention contest can be an extended brainstorming session that lasts for several days or weeks. Several teams compete with each other to come up with the best solutions (which are often inventions) to one or two problems. By offering prizes and recognition, the invention teams work hard to develop valuable solutions for the company.

Try an invention contest or a fun brainstorming activity with your team and test the results.

Review Chapter 7 for more ideas.

7. Failing to Identify Inventions Before Public Disclosure

Don't accidentally give away your valuable inventions by publicly disclosing the inventions before they are properly protected. If you perform certain types of public disclosure, such as trade shows, white paper publications, and public sales presentations, you may lose your right to file a patent application in foreign countries. Specific types of activities, such as offering to sell a product that contains an invention, won't prevent you from filing a patent application covering the invention in the United States, but those activities trigger a one-year deadline within the United States for filing a patent application. For these activities, you still have the right to file a US patent application, but you must file the application within one year of the triggering event.

Let everyone in your organization know about the potential problems associated with public disclosures, and implement a system to review and approve all planned public disclosures that may reveal invention details.

Review Chapter 8 for more information about important events that require patent action.

Most companies seem to have stumbled into several of these pitfalls. Now that you're aware of these potential mistakes, start taking steps right now to avoid them. If you already made one of these mistakes, begin correcting the situation today.

CONCLUSION

"An explicit innovation strategy helps you design a system to match your specific competitive needs." Harvard Business Review

Now you know my three-step system for identifying, evaluating, and protecting your critical inventions. If you were confused about the patent process before reading this book, I hope you realize it's not as complicated as you might have thought. Yes, there are multiple activities for each of the three steps, but there is a definite system. If you follow this system, you can enjoy peace of mind knowing your inventions are secure and enjoying a competitive advantage in your industry.

A strong patent portfolio is a critical asset for every technology company. Carefully selecting high-value inventions for your portfolio improves your company valuation and positions the business as an industry leader.

Follow the three steps described in this book, and enjoy the same benefits experienced by hundreds of other companies that follow the same system. Don't waste more time and money on mediocre patents. Your patent portfolio deserves the best inventions your team has created.

This is the same system I use every day with my own clients. As you can tell, I am passionate about this topic. That's why I wrote this book; it lets me easily share this system with hundreds (hopefully thousands) of technology companies. I love seeing tech companies grow and produce new inventions that change the world. I hope this book stimulates the development of more inventions that will help you and other tech companies succeed by changing more lives.

Start implementing this system in your own business today, and begin growing (or expanding) your own patent portfolio filled with industry-leading inventions.

Please share your success stories with me!

ABOUT THE AUTHOR
WHO I AM

Since I was five years old, I have enjoyed learning how things work. I also love creating things – from Lego® spaceships to designing computer systems.

When I was a kid, I liked to take things apart to learn how they worked. Then, I would try to put them back together to better understand how they operated. Usually, I was given broken appliances and other items to disassemble. One day, when I had run out of broken items, I took apart my mom's brand-new vacuum cleaner; she was not happy about that! Fortunately, my dad was an engineer, so he saved me and reassembled the vacuum cleaner (and it still worked).

My passion for learning new things led me to work as an electrical engineer and an intellectual property attorney. Today, I help technology companies of all sizes foster and protect valuable inventions.

During the past 20 years, I am grateful to have worked with hundreds of technology companies and more than 1000 inventors. I've had fun working with autonomous vehicles, video games, robots, and artificial intelligence. I have worked with all types of businesses – from small startups to large corporations, including Intel, HP, Microsoft, and eBay. I also enjoy writing about intellectual property and innovation. My articles have been featured in *Entrepreneur*, *The Huffington Post*, and other publications.

I love being part of the creative process and helping companies thrive in their marketplace by identifying, developing, and protecting cutting-edge inventions. A great day for me includes brainstorming with a group of inventors and business leaders to discuss the exciting technology changes coming in the next few years.

The three-step system described in this book was developed over the past 20 years of working with numerous technology companies. This is the same system I use today with all of my clients. You can use it in your own business to identify and secure your most important inventions.

NEXT STEPS

I hope this book has sparked some new thinking and innovative ideas for you. Whenever you're ready, here are a couple of ways I can help you implement my three-step system in your own business.

1. Subscribe to the Tech Leader Talk podcast. TechLeaderTalk.com

2. Work with me to build a strong and secure patent portfolio that protects your critical inventions, provides a competitive advantage, and increases the value of your company. To find out if we're a fit, email me at steve@stevesponseller.com.

STEVEN SPONSELLER

RESOURCES

Here are some free invention tools to help you identify and protect your most valuable inventions:

- Invention Disclosure Form template to capture your critical inventions

- Invention Evaluation Checklist template to prioritize your inventions

- One-Page Infographic summarizing the three-step system described in this book

Visit SteveSponseller.com/tools to get all of these resources.

www.ingramcontent.com/pod-product-compliance
Lightning Source LLC
Chambersburg PA
CBHW031439210526
45464CB00005B/2268